Some Notes on the Theory of Hilbert Spaces
of Analytic Functions of the Unit Disc

by
Jorge-Nuno O. Silva

ISBN: 1-58112-023-0

DISSERTATION.COM

1998

Some Notes on the Theory of Hilbert Spaces of Analytic Functions on the Unit Disc

by

Jorge-Nuno O. Silva

M. A. (University of California at Berkeley) 1991

A dissertation submitted in partial satisfaction of the
requirements for the degree of
Doctor of Philosophy
in

Mathematics

in the
GRADUATE DIVISION
of the
UNIVERSITY of CALIFORNIA at BERKELEY

Committee in charge:
        Professor Donald Sarason, Chair
        Professor William Arveson
        Professor Steven Evans

1994

Copyright
by

Some Notes on the Theory of Hilbert Spaces of Analytic Functions on the Unit Disc
© copyright 1994
by
Jorge-Nuno O. Silva

Acknowledgements

Several people helped me in the process of coming to Berkeley and living here. Among them Tone Feijó, Carlos Sarrico and Jonathan Walden. I thank them very much.

I thank JNICT-Programa CIÊNCIA and Fundação Luso-Americana para o Desenvolvimento for their finantial support.

I am grateful to Professor D. Sarason for his help and advice.

To José Luis Fachada, João Santos Guerreiro and Nuno Costa Pereira, after all these years, my thanks.

I thank my family – Laura, Manuel and Raquel – for their love and support.

To
Laura and Manuel.

Table of Contents

Acknowledgements, iii

Dedication, iv

Table of Contents, v

Chapter 1, 1

Chapter 2, 11

Chapter 3, 16

Bibliography, 21

Introduction

Let $\mathbb{D}$ be the open unit disk in $\mathbb{C}$. This work deals with some aspects of the theory of Hilbert spaces of analytic functions on $\mathbb{D}$.

H^2 is defined to be the set of holomorphic functions on $\mathbb{D}$ with square summable Maclaurin coefficients. The norm of an element f of H^2 can be given by a series, an integral over $\partial \mathbb{D}$ or an area integral over $\mathbb{D}$:

$$\|f\|_{H^2}^2 = \sum_{n \geq 0} |\hat{f}(n)|^2 \qquad (0.1)$$

$$\|f\|_{H^2}^2 = \frac{1}{2\pi} \int_0^{2\pi} |f(e^{i\theta})|^2 d\theta \qquad (0.2)$$

$$\|f\|_{H^2}^2 = |f(0)|^2 + 2\int_{\mathbb{D}} |f'(z)|^2 \log \frac{1}{|z|} dA(z) \qquad (0.3)$$

where $d\theta$ is the Lebesgue measure on $\partial \mathbb{D}$, and dA is the normalized Lebesgue measure on $\mathbb{D}$, $dA(re^{i\theta}) = (1/\pi)rd\theta dr$. The values on $\partial \mathbb{D}$ are the radial limits:

$$f(e^{i\theta}) = \lim_{r \to 1} f(re^{i\theta})$$

See [5, 6].

An important subset of H^2 is the Dirichlet space, D, of the holomorphic functions on $\mathbb{D}$ with

$$\int_{\mathbb{D}} |f'(z)|^2 dA(z) < \infty ; \qquad (0.4)$$

the norm of f is given by

$$\|f\|_D^2 = |f(0)|^2 + \int_{\mathbb{D}} |f'(z)|^2 dA(z) \cdot \qquad (0.5)$$

In [9] Richter introduced a family of spaces between the Dirichlet space D and Hardy space H^2 which we briefly describe now. Let μ be a finite positive Borel measure on $\partial \mathbb{D}$. Define the harmonic function ϕ_μ on $\mathbb{D}$ by

$$\phi_\mu(z) = \int_{\partial \mathbb{D}} \frac{1-|z|^2}{|\zeta - z|^2} d\mu(\zeta). \qquad (0.6)$$

If $\mu = 0$ define $D(\mu) = H^2$; otherwise $D(\mu)$ is defined to be the space of analytic functions on $\mathbb{D}$ with

$$\int_{\mathbb{D}} |f'(z)|^2 \phi_\mu(z) dA(z) < \infty; \qquad (0.7)$$

$D(\mu)$ is a Hilbert space with norm given by

$$\|f\|_{D(\mu)}^2 = \|f\|_{H^2}^2 + \int_{\mathbb{D}} |f'(z)|^2 \phi_\mu(z) dA(z). \qquad (0.8)$$

We now describe another technique to produce subspaces of H^2; this one is a particular case of a construction in [3], see also [10]. Let $A : H^2 \to H^2$ be a bounded linear operator. As a set $M(A)$ is the range of A. The norm is defined by

$$\|A(x)\|_{M(A)} = \|x\| \text{ for all } x \perp \ker(A) \qquad (0.9)$$

In [9] it is shown that $D(\delta_\zeta) = M(\overline{\zeta} - S^*)$ where S is the unilateral shift on H^2. The equivalence of the norms carried by those spaces will permit us to define an operator $A \in L(H^2)$ satisfying

$$\|A(f)\|_{H^2} = \|f\|_{D(\delta_1)}$$

which we'll study in some detail. It will be shown that it is a rank one perturbation of a Toeplitz operator with real valued symbol.

These results will be generalized to operators of the form $\prod_{i=1}^n (\overline{\zeta_i} - S^*)$ where $\zeta_i \in \partial \mathbb{D}$ for $i = 1, \cdots, n$. We'll show that

$$M\left(\prod_{i=1}^n (\overline{\zeta_i} - S^*)\right) = D\left(\sum_1^n \delta_{\zeta_i}\right)$$

and that the associated operator A is a rank n perturbation of a Toeplitz operator. We will perform a similar analysis for for the space $M((1-S^*)^2)$.

Douglas, in [4], used a kind of integral Richter in [8] called a local Dirichlet integral. For a function f and $\zeta \in \partial \mathbb{D}$, it is denoted by $D_\zeta(f)$, and has the following definition:

$$D_\zeta(f) = \frac{1}{2\pi} \int_0^{2\pi} \left|\frac{f(e^{i\theta}) - f(\zeta)}{e^{i\theta} - \zeta}\right|^2 d\theta. \qquad (0.10)$$

Introduction

Richter proved that $D(\mu)$ is the space of functions f with $D_\zeta(f) \in L^1(\partial \mathbb{D}, \mu)$. In Chapter 2 we'll use a similar integral

$$D_\zeta^{(2)}(f) = \frac{1}{2\pi} \int_0^{2\pi} \left| \frac{e^{i\theta} f(e^{i\theta}) - e^{i\theta}(f(\zeta) + f'(\zeta)) + f'(\zeta)}{(e^{i\theta} - \zeta)^2} \right|^2 d\theta,$$

to characterize another space of analytic functions on $\mathbb{D}$. We'll show that

$$\int_{\mathbb{D}} |f''(z)|^2 \frac{1-|z|^2}{|1-z|^2} dA(z) < \infty \text{ iff } D_\zeta^{(2)}(f) < \infty.$$

Chapter 3 deals with composition operators. If ϕ is a holomorphic self map of $\mathbb{D}$ the composition operator induced by ϕ is defined by

$$C_\phi(f)(z) = f((\phi(z))$$

It is a consequence of classical results of Littlewood [7] that the composition operator on H^2 associated with such a ϕ is bounded. Shapiro gave a complete characterization of the compact ones (see [11, 12].) We will obtain similar results for the space $D(\delta_\zeta)$.

Chapter 1

Let $\zeta \in \partial \mathbb{D}$. Richter in [8] showed that

$$\int_{\mathbb{D}} |f'(z)|^2 \frac{1-|z|^2}{|\zeta-z|^2} dA(z) = D_\zeta(f) = \frac{1}{2\pi} \int_0^{2\pi} \left| \frac{f(e^{i\theta}) - f(\zeta)}{e^{i\theta} - \zeta} \right|^2 d\theta \tag{1.1}$$

so

$$\|f\|^2_{D(\delta_\zeta)} = \|f\|^2_{H^2} + D_\zeta(f) \tag{1.2}$$

In [8] it was also established that $D(\delta_\zeta) = M(\overline{\zeta} - S^*)$. The continuity of the identity operator $i: M(\overline{\zeta} - S^*) \to D(\delta_\zeta)$ together with the Open Mapping Theorem give that the norms carried by those spaces are equivalent. Our first result gives quantitative information about the relation between them.

Theorem 1 *The following sharp bounds hold for $f \in D(\delta_\zeta)$:*

$$\frac{1}{\sqrt{2}} \leq \frac{\|f\|_{D(\delta_\zeta)}}{\|f\|_{M(\overline{\zeta}-S^*)}} < \sqrt{5}. \tag{1.3}$$

Proof: Without loss of generality take $\zeta = 1$. We have

$$M(1 - S^*) = D(\delta_1) = (z-1)H^2 + \mathbb{C}. \tag{1.4}$$

For $g \in H^2$ and $f = (z-1)g + c \in M(1-S^*)$ a calculation gives

$$\|f\|^2_{M(1-S^*)} = \|g\|^2_{H^2} + |c|^2$$

and

$$\|f\|^2_{D(\delta_1)} = 3\|g\|^2_{H^2} + |c|^2 - \Re \overline{c} g(0) - 2\Re <zg, g>,$$

so

$$1 + 2\frac{\|g\|_{H^2}^2 - \Re<zg,g> - |cg(0)|}{\|g\|_{H^2}^2 + |c|^2} \le$$

$$\frac{\|f\|_{D(\delta_1)}^2}{\|f\|_{M(1-S^*)}^2} \le \qquad (1.5)$$

$$1 + 2\frac{\|g\|_{H^2}^2 - \Re<zg,g> + |cg(0)|}{\|g\|_{H^2}^2 + |c|^2}.$$

Now we have

$$1 + 2\frac{\|g\|_{H^2}^2 - \Re<zg,g> + |cg(0)|}{\|g\|_{H^2}^2 + |c|^2} < 5 - \frac{2|cg(0)|}{\|g\|_{H^2}^2 + |c|^2} \le 5.$$

For fixed g, studying the rightmost expression in (1.5) as a function of $|c|$ we conclude that it is at most

$$1 + \frac{1}{\sqrt{(\frac{\|g\|^2 - \Re<zg,g>}{|g(0)|^2})^2 + \frac{\|g\|^2}{|g(0)|^2}} - \frac{\|g\|^2 - \Re<zg,g>}{|g(0)|^2}}. \qquad (1.6)$$

Let h be defined by

$$h(z) = \sum_{i=0}^{n} (-1)^i z^i.$$

Using h in the place of g in (1.6) gives

$$1 + \frac{1}{\sqrt{(2n+1)^2 + n + 1} - (2n+1)}$$

which increases to 5 as $n \to \infty$, so the constant $\sqrt{5}$ is sharp in (1.3).

A similar computation shows that the leftmost expression in (1.5) is at least

$$1 - \frac{1}{\sqrt{(\frac{\|g\|^2 - \Re<zg,g>}{|g(0)|^2})^2 + \frac{\|g\|^2}{|g(0)|^2}} + \frac{\|g\|^2 - \Re<zg,g>}{|g(0)|^2}} \qquad (1.7)$$

and we get the lower bound in (1.3).

Let $h_{1/2}$ be given by

$$h_{1/2}(z) = \sum_{i=0}^{\infty} 2^{-i} z^i.$$

We get 1/2 if we substitute $h_{1/2}$ for g in (1.7), which shows that the constant in the left inequality in (1.3) is also sharp. □

This equivalence of norms permits us to define a new norm in H^2, $\|\ \|_\zeta$, by

$$\|f\|_\zeta = \|(\overline{\zeta} - S^*)f\|_{D(\delta_\zeta)}$$

Let $<,>_\zeta$ be the corresponding inner product. The Riesz Representation Theorem gives the existence of an operator $A \in L(H^2)$ such that

$$<f,g>_\zeta = <f, A(g)>_{H^2} \quad \forall f, g \in H^2$$

A calculation with the polarization identity shows that we have

$$A(1) = 1 - \overline{\zeta}z, \quad A(z^n) = -\zeta z^{n-1} + 3z^n - \overline{\zeta}z^{n+1} \text{ for } n \geq 1.$$

So we have

Theorem 2 *Let A be as above. Then A is the sum of a Toeplitz operator, T_ϕ, with a rank one operator, B, where*

$$\phi(z) = -\zeta\overline{z} + 3 - \overline{\zeta}z \text{ and } B\left(\sum_{i=0}^{\infty} c_i z^i\right) = -2c_0$$

When the measure μ is a linear combination of Dirac point masses we have the following results.

Theorem 3 *Let $\zeta_1, \cdots, \zeta_n$ be n distinct elements of $\partial \mathbb{D}$. Then*

a) $$M\left(\prod_{i=1}^{n}(\overline{\zeta}_i - S^*)\right) = D\left(\sum_{i=1}^{n} \delta_{\zeta_i}\right) \quad (1.8)$$

with equivalence of norms.

b) *If we define a new norm in H^2 by*

$$\|g\|_n = \|\prod_{1}^{n}(\overline{\zeta}_i - S^*)g\|_{D(\sum_1^n \delta_{\zeta_i})} \quad (1.9)$$

for

$$g(z) = \sum_{i=0}^{\infty} c_i z^i$$

Chapter One 7

then

$$\|g\|_n^2 = \sum_{i\geq 0}\left|\sum_{j=0}^{n}\sigma_{n-j}^{n}c_{i+j}(-1)^j\right|^2 + \sum_{i\geq 1}{\sum}'\left|\sum_{j=0}^{n-1}\sigma_{n-1-j}^{n-1}c_{i+j}(-1)^j\right|^2 \qquad (1.10)$$

where for $m \geq 1$ *we have*

$$\sigma_k^m = 0 \text{ for } k<0, \sigma_0^m = 1, \sigma_1^m = \sum_{i=1}^{m}\overline{\zeta_i}, \sigma_2^m = \sum_{i<j}\overline{\zeta_i\zeta_j}, \cdots, \sigma_m^m = \prod_{i=1}^{m}\overline{\zeta_i} \qquad (1.11)$$

are the elementary symmetric polynomials on $\overline{\zeta_1},\cdots,\overline{\zeta_n}$ *and* ${\sum}'$ *is the sum over the sets*

$$\{\overline{\zeta_1},\cdots,\overline{\zeta_{n-1}}\}, \{\overline{\zeta_1},\cdots,\overline{\zeta_{n-2}},\overline{\zeta_n}\},\cdots,\{\overline{\zeta_2},\cdots,\overline{\zeta_n}\}. \qquad (1.12)$$

Proof: Let $n = 2$. We have, for $f = (\overline{\zeta}_1 - S^*)(\overline{\zeta}_2 - S^*)g$,

$$f(z) = \overline{\zeta}_1\overline{\zeta}_2 g(0) - (\overline{\zeta}_1+\overline{\zeta}_2)g'(0) + \overline{\zeta}_1\overline{\zeta}_2 zg'(0) + (\overline{\zeta}_1 - \overline{z})(\overline{\zeta}_2 - \overline{z})(g(z) - g(0) - zg'(0))$$

hence

$$M((\overline{\zeta}_1 - S^*)(\overline{\zeta}_2 - S^*)) = (\zeta_1 - z)(\zeta_2 - z)\mathrm{H}^2 + z\mathbb{C} + \mathbb{C} \qquad (1.13)$$

and the inclusion $M((\overline{\zeta}_1 - S^*)(\overline{\zeta}_2 - S^*)) \subset \mathrm{D}(\delta_{\zeta_1} + \delta_{\zeta_2})$ follows.

Let $f \in \mathrm{D}(\delta_{\zeta_1} + \delta_{\zeta_2}) = \mathrm{D}(\delta_{\zeta_1}) \cap \mathrm{D}(\delta_{\zeta_2})$. Then, by (1.4), there are constants k_1, k_2 such that

$$\frac{f(z) - k_1}{z - \zeta_1}, \frac{f(z) - k_2}{z - \zeta_2} \in \mathrm{H}^2.$$

The difference of these functions is also in H^2, so, by (1.13), we get

$$\mathrm{D}(\delta_{\zeta_1} + \delta_{\zeta_2}) \subset \mathrm{M}((\overline{\zeta}_1 - \mathrm{S}^*)(\overline{\zeta}_2 - \mathrm{S}^*)).$$

The equivalence of the norms carried by these spaces is a consequence of the continuity of the identity and the Open Mapping Theorem.

The proof for general $n > 2$ follows the same lines. □

Let $<.,.>_n$ be the inner product associated with $\|.\|_n$. As before, we may define an operator $A \in L(\mathrm{H}^2)$ by $<f,g>_n = <f,A(g)>$ for $f,g \in \mathrm{H}^2$. It follows from (1.10) that $A = T_\phi + B$ with

$$\phi(z) = a(0,n)\bar{z}^n + a(1,n)\bar{z}^{n-1} + \cdots + a(n,n) + a(n+1,n)z + \cdots + a(2n,n)z^n$$

where for $0 \le k \le m \le n$

$$\overline{a(k,m)} = (-1)^{m-k} \left(\sum_{j=n-m}^{n-m+k} \sigma^n_{m-k+j} \overline{\sigma^n_j} + {\sum_{j=n-m}^{n-1-m+k}}' \sigma^{n-1}_{m-k+j} \overline{\sigma^{n-1}_j} \right)$$

and $a(m,k) = \overline{a(k,m)}$. Note that for $s > 0$, $a(n+s,n) = \overline{a(n-s,n)}$, so ϕ is real valued. B has the hermitic matrix representation given by

$$B_{ij} = \begin{cases} a(j-1, i-1) - a(n, n+i-j) & \text{if } 1 \le j \le i \le n \\ 0 & \text{if } i > n \text{ or } j > n \end{cases}$$

Example 1 *For $n = 2$ we get*

$$\phi(z) = \zeta_1 \zeta_2 \bar{z}^2 - 3(\overline{\zeta_1 + \zeta_2})\bar{z} + (6 + |\zeta_1 + \zeta_2|^2) - 3(\overline{\zeta_1 + \zeta_2})z + \overline{\zeta_1\zeta_2}z^2$$

and to B corresponds the matrix

$$\begin{pmatrix} -5 - |\zeta_1+\zeta_2|^2 & 2(\zeta_1+\zeta_2) & 0 & \cdots \\ 2(\overline{\zeta_1+\zeta_2}) & -3 & 0 & \cdots \\ 0 & 0 & 0 & \cdots \\ \vdots & \vdots & \vdots & \ddots \end{pmatrix}.$$

Now we do a particular choice of the points in the circle which leads to some simplifications.

Corollary 1 *Let $\zeta_1, \cdots, \zeta_n$ be the n-roots of 1, $n \ge 1$. Then $A = T_\phi + B$ where*
a)

$$\phi(z) = -\bar{z}^n + 2 + n^2 - z^n$$

and

$$B = diag(-1 - n^2, -1 + n - n^2, \cdots, -1 - n).$$

Chapter One

b) The eigenvalues of A are

$$\lambda(k) = 2 + n^2 - \frac{1+(1+n^2-kn)^2}{1+n^2-kn} \quad \text{for } k = 0, \cdots, n-1.$$

The eigenspace associated with each $\lambda(k)$ is spanned by $f(z) = \sum_{i=0}^{\infty} c_i z^i$ with

$$c_i = \begin{cases} (1+kn-\lambda(k))^s & \text{if } i = ns+k \\ 0 & \text{if } i \not\equiv k \ (mod \ n) \end{cases}$$

Proof:

a) It is enough to notice that under the hypothesis we have

$$|\sigma_0^n| = |\sigma_n^n| = |\sigma_k^{n-1}| = 1 \text{ for } k = 0, \cdots, n-1$$

and

$$\sigma_k^n = 0 \quad \text{for } k = 1, \cdots, n-1.$$

b)

Let $f(z) = \sum_{i=0}^{\infty} c_i z^i$ be an eigenvector relative to the eigenvalue λ. Then we have the system

$$c_{n+k} = (1+kn-\lambda)c_k \qquad \text{for } 0 \le k \le n-1$$
$$c_k = (2+n^2-\lambda))c_{k-n} - c_{k-2n} \quad \text{for } k \ge 2n$$

and the result follows. $\square$

We now study the space $M\left((1-S^*)^2\right)$.

Let the subspace of H^2, D_2, be defined by

$$D_2 = \left\{ f \in H^2 : \frac{f(z) - f(1) - f'(1)(z-1)}{(z-1)^2} \in H^2 \right\}.$$

with norm $\|.\|_{D_2}$ given by

$$\|f\|_{D_2}^2 = \|f\|_{H^2}^2 + \left\|\frac{f(z) - f(1) - f'(1)(z-1)}{(z-1)^2}\right\|_{H^2}^2.$$

It is a consequence of Richter's result $M(1-S^*) = D(\delta_1)$ that we have $M\left((1-s^*)^2\right) = D_2$, also with equivalence of norms.

Let $f = (1-S^*)^2 g$ for some $g \in H^2$. Then

Chapter One

and
$$\|f\|^2_{M((1-S^*)^2)} = |g(0)|^2 + |(S^*g)(0)|^2 + \|(S^*)^2 g\|^2_{H^2}$$

$$\|f\|^2_{D_2} = \|f\|^2_{H^2} + \|(S^*)^2 g\|^2_{H^2}.$$

If we define in H^2 a new norm by

$$\|g\|_2 = \|(1-S^*)^2 g\|_{D_2}$$

and if the corresponding inner product is denoted by $<.,.>_2$, an operator $A \in L(H^2)$ must exist such that

$$<f,g>_2 = <f, Ag> \quad \forall f, g \in H^2.$$

A calculation gives

$$A1 = 1 - 2z - z^2$$
$$Az = -2 + 5z - z^3$$
$$Az^k = -z^{k-2} + 7z^k - z^{k+2} \quad \text{for } k \geq 2$$

so $A = T_\phi + B$ where $\phi(z) = -\bar{z}^2 + 7 - z^2$ and B has the matrix representation

$$\begin{pmatrix} -6 & -2 & 0 & \cdots \\ -2 & -2 & 0 & \cdots \\ 0 & 0 & 0 & \cdots \\ \vdots & \vdots & \vdots & \ddots \end{pmatrix}.$$

We have $sp_e(A) = sp(T_\phi) = [5,9]$, $A > 0$, $<A1, 1> = 1$ so A must have an eigenvalue in $(0,1)$, λ, say. Let $f(z) = \sum_{i=0}^{\infty} c_i z^i$ be an eigenvector associated with λ; then

$$c_0 - 2c_1 - c_2 = \lambda c_0$$
$$-2c_0 + 5c_1 - c_3 = \lambda c_1$$
$$-c_{n-2} + 7c_n - c_{n+2} = \lambda c_n \quad \text{for } n \geq 2$$

which gives

$$\lambda = \frac{5}{2} - \frac{7}{4}\sqrt{2} \text{ and } f(z) = \frac{1 + \left(\sqrt{2} - 1\right)z}{1 - \left(\frac{1}{2} - \frac{\sqrt{5}}{4}\right)z^2}.$$

Chapter 2

We introduce another Hilbert space of analytic functions on $\mathbb{D}$, $D^{(2)}(\delta_1)$, by

$$D^{(2)}(\delta_1) = \left\{ f \in H^2 : \int_{\mathbb{D}} |f''(z)|^2 \frac{1-|z|^2}{|1-z|^2} dA(z) < \infty \right\}$$

with the norm

$$\|f\|^2_{D^{(2)}(\delta_1)} = \|f\|^2_{H^2} + \int_{\mathbb{D}} |f''(z)|^2 \frac{1-|z|^2}{|1-z|^2} dA(z).$$

A similar definition could be given with any $\zeta \in \partial \mathbb{D}$ in the place of 1.

For $f \in H^1$ and $\alpha, \beta \in \mathbb{C}$ let

$$I_1(f, \alpha, \beta) = \int_0^{2\pi} \left| \frac{e^{i\theta} f(e^{i\theta}) - e^{i\theta}\alpha + \beta}{(e^{i\theta} - 1)^2} \right|^2 d\theta.$$

It is clear that, for a fixed f, $I_1(f, \alpha, \beta)$ can be finite for at most one choice of the constants α, β. Our next result characterizes those values.

Theorem 1 *Let $f \in H^1$ with $I_1(f, \alpha, \beta) < \infty$ for some constants α, β. Then $\alpha = f(1) + f'(1)$ and $\beta = f'(1)$.*

By $f'(1)$ we are referring to the radial limit of f' at 1, which is the same as the angular derivative of f at 1, see [2].

For the proof we need a well known result that we state without proof

Lemma 1 *Let $g \in H^2$; then*

$$\lim_{|z| \to 1} (1-|z|^2)|g(z)|^2 = 0. \tag{2.1}$$

Chapter Two

Proof of Theorem 1: The hypothesis means that

$$g(z) := \frac{zf(z) - \alpha z + \beta}{(z-1)^2} \in H^2$$

We have

$$|zf(z) - \alpha z + \beta|^2 = \left|(z-1)^2 g(z)\right|^2 = \frac{|z-1|^4}{1-|z|^2}(1-|z|^2)|g(z)|^2$$

so, by the previous lemma,

$$\lim |zf(z) - \alpha z + \beta|^2 = 0$$

when z approaches 1 within any region

$$\left\{z \in \mathbb{C} : |z-1|^4 < k(1-|z|^2)\right\}.$$

In particular the nontangential limit of $|zf(z) - \alpha z + \beta|^2$ at 1 is 0, so $f(1) = \alpha - \beta$.

Similarly,

$$\left|\frac{zf(z) - \alpha z + \beta}{z-1}\right|^2 = |(z-1)^2 g(z)|^2 = \frac{|z-1|^2}{1-|z|^2}(1-|z|^2)|g(z)|^2$$

so

$$\lim \left|\frac{zf(z) - \alpha z + \beta}{z-1}\right|^2 = 0 \qquad (2.2)$$

when z approaches 1 from within any oricyclic region

$$\left\{z \in \mathbb{C} : |z-1|^2 < k(1-|z|^2)\right\},$$

which implies that the nontangential limit of the expression in (2.2) is 0, and this fact gives $\alpha = f(1) + f'(1)$. □

This result makes our next definition meaningful.

Definition 1 *If, for $f \in H^1$, no choice of α, β produces $I_1(f, \alpha, \beta)$ finite, put $D_1^{(2)}(f) = \infty$; otherwise*

$$D_1^{(2)}(f) = \frac{1}{2\pi}\int_0^{2\pi}\left|\frac{e^{i\theta}f(e^{i\theta}) - e^{i\theta}(f(1) + f'(1)) + f'(1)}{(e^{i\theta}-1)^2}\right|^2 d\theta.$$

We now start characterizing the space $D^{(2)}(\delta_1)$.

Chapter Two

Theorem 2 *The polynomials are dense in* $D^{(2)}(\delta_1)$.

Proof: Let $f \in D^{(2)}(\delta_1)$. Then $f' \in D(\delta_1)$. As the polynomials are dense in $D(\delta_1)$, there exists a sequence of polynomials, (p_n), such that $p_n \to f'$ in $D(\delta_1)$. Then

$$\lim_{n\to\infty}\left(\|p_n - f'\|_{H^2}^2 + \int_{\mathbb{D}} |p'_n(z) - f''(z)|^2 \frac{1-|z|^2}{|1-z|^2} dA(z)\right) = 0$$

so

$$\|q_n - f\|_{D^{(2)}(\delta_1)} = o(1) \quad (n \to \infty)$$

for $q_n = f(0) + \int p_n$. $\square$

Before the main result of this chapter we need the following

Lemma 2 *Let* $n \geq m$ *be natural numbers. We have*

a)

$$\frac{1}{2\pi}\int_0^{2\pi} \frac{e^{i(n+1)\theta} - e^{i\theta}(n+1) + n}{(e^{i\theta}-1)^2} \cdot \frac{e^{-i(m+1)\theta} - e^{-i\theta}(m+1) + m}{(e^{-i\theta}-1)^2} d\theta =$$

$$= \frac{m(m+1)(3n-m+1)}{6}$$

b)

$$\int_{\mathbb{D}} (z^n)''\overline{(z^m)''}\frac{1-|z|^2}{|1-z|^2}dA(z) = n(m+1)(m-1) \text{ for } m \geq 2$$

c) *There are positive constants* c_1, c_2 *such that*

$$c_1\int_{\mathbb{D}} |p(z)''|^2 \frac{1-|z|^2}{|1-z|^2}dA(z) \leq D_1^{(2)}(p) \leq c_2\int_{\mathbb{D}} |p(z)''|^2 \frac{1-|z|^2}{|1-z|^2}dA(z)$$

for every polynomial p *($p(0) = p'(0) = 0$ is assumed in the rightmost inequality.)*

Proof:

a) Note that for any natural k we have

$$\frac{z^{k+1} - z(k+1) + k}{(z-1)^2} = \sum_{i=1}^{k} iz^{k-i}$$

so

$$\frac{1}{2\pi}\int_0^{2\pi} \frac{e^{i(n+1)\theta} - e^{i\theta}(n+1) + n}{(e^{i\theta} - 1)^2} \cdot \frac{e^{-i(m+1)\theta} - e^{-i\theta}(m+1) + m}{(e^{-i\theta} - 1)^2} d\theta =$$

$$= \sum_{i=1}^m (n - m + i)i$$

and a) follows.

b) Using polar coordinates we have

$$\int_{\mathbb{D}} (z^n)''\overline{(z^m)''} \frac{1 - |z|^2}{|1 - z|^2} dA(z) =$$

$$2n(n-1)m(m-1) \int_0^1 r^{n+m-3} \frac{1}{2\pi} \int_0^{2\pi} e^{i(n-m)\theta} \frac{1-r^2}{|e^{i\theta} - r|^2} d\theta dr$$

and b) follows.

c) is a simple consequence of a) and b). □

Theorem 3 *Let* $f \in H^2$. *Then*

$$D_1^{(2)}(f) < \infty \text{ iff } f \in D^{(2)}(\delta_1) \tag{2.3}$$

Proof: Suppose $D_1^{(2)}(f) < \infty$. Then

$$g(z) := \frac{zf(z) - z(f(1) + f'(1)) + f'(1)}{(z-1)^2} \in H^2$$

So there exists a sequence of polynomials (q_n) such that $q_n \to g$ in H^2. As $g(0) = f'(1)$ we can assume that these polynomials satisfy $q_n(0) = f'(1)$.

For each n let

$$p_n(z) := \frac{q_n(z)(z-1)^2 + z(f(1) + f'(1)) + f'(1)}{z}$$

So p_n is a polynomial for each n. We have

$$D_1^{(2)}(p_n - f) = \|q_n - g\|_{H^2}^2 = o(1) \ (n \to \infty)$$

Then, using the previous Lemma, we conclude that (p_n) is a Cauchy sequence in $D^{(2)}(\delta_1)$. So p_n converges for some h in $D^{(2)}(\delta_1)$. But p_n clearly converges pointwise to f in $\mathbb{D}$, so $h = f$.

Chapter Two

Suppose now that $f \in D^{(2)}(\delta_1)$. As (2.3) is trivially satisfied by all linear polynomials, we may assume $f(0) = f'(0) = 0$. Lemma 1 garantees the existence of a sequence of polynomials, p_n, converging to f in $D^{(2)}(\delta_1)$. We can assume also that $p_n(0) = p'_n(0) = 0$. Let q_n be defined on $\mathbb{D}$ by

$$q_n(z) = \frac{zp_n(z) - z(p_n(1) + p'_n(1)) + p'_n}{(z-1)^2}$$

The previous Lemma shows that q_n is a polynomial for each n. Part c) of Lemma 2 gives that (q_n) is a Cauchy sequence in H^2, so $q_n \to g \in H^2$.

Pointwise we have $p_n \to f$, $q_n \to g$ so $p_n(1)$, $p'_n(1)$ converge to x, y, respectively, say.

Hence

$$g(z) = \frac{zf(z) - z(x+y) + y}{(z-1)^2} \in H^2$$

so $D_1^{(2)}(f) < \infty$. □

Chapter 3

In this chapter we study the action of composition operators on $D(\delta_1)$. We find convenient to use another norm in this space.

Lemma 1 *The following norms are equivalent on $D(\delta_1)$.*

$$\|f\|^2_{D(\delta_1)} = \|f\|^2_{H^2} + \int_{\mathbb{D}} |f'(z)|^2 \frac{1-|z|^2}{|1-z|^2} dA(z)$$

$$\|f\|^2_\bullet = |f(0)|^2 + \int_{\mathbb{D}} |f'(z)|^2 \frac{1-|z|^2}{|1-z|^2} dA(z).$$

Proof: This is an easy consequence of the fact that the usual norm on H^2 is equivalent to

$$\|f\|^2 = |f(0)|^2 + \int_{\mathbb{D}} |f'(z)|^2 (1-|z|^2) dA(z)$$

see [6]. □

We'll need also two kinds of auxiliary functions, which we now introduce. Let $0 \neq p \in \mathbb{D}$; the automorphism of $\mathbb{D}$ that sends p to the origin is

$$\alpha_p(z) = \frac{p-z}{1-\bar{p}z};$$

define f_p on $\mathbb{D}$ by

$$f_p(z) = \frac{|1-p|\sqrt{1-|p|^2}}{|p|(1-\bar{p}z)}.$$

We now summarize the properties of these functions we'll need.

Lemma 2 *a) $\{f_p : p \in \mathbb{D}\}$ is a bounded family in $D(\delta_1)$.*

b) $f_p \to 0$ as $|p| \to 1$ uniformly on compacta on $\mathbb{D}$.

Chapter Three

c) We have the following identity for $z \in \mathbb{D}$

$$|f_p'(z)|^2 = \frac{|1-p|^2}{1-|p|^2}|\alpha_p'(z)|^2 .$$

Proof:

a) A calculation shows that

$$\|f_p\|_\bullet^2 \le 3 \quad \forall z \in \mathbb{D}.$$

b) Let $0 < r < 1$. Then, on $r\mathbb{D}$, we have

$$|f_p(z)| < \frac{|1-p|\sqrt{1-|p|^2}}{|p|(1-r|p|)} ,$$

and b) follows.

A simple manipulation gives c). □

We now state a result to help us estimate the norm of a composition of functions. It is a simple consequence of the Non-Univalent Change of Variables Formula, see [1].

Theorem 1 *Let ϕ be a holomorphic self map of $\mathbb{D}$. Then, for $f \in \mathrm{D}(\delta_1)$, we have*

$$\|C_\phi(f)\|_\bullet^2 = \|f \circ \phi\|_\bullet^2 = |f(\phi(0))|^2 + \int_{\phi(\mathbb{D})} |f'(w)|^2 M_\phi(w) dA(w) \tag{3.1}$$

where

$$M_\phi(w) = \sum_{\phi(z)=w} \frac{1-|z|^2}{|1-z|^2} . \tag{3.2}$$

We now characterize the bounded composition operators on $\mathrm{D}(\delta_1)$.

Theorem 2 *Let ϕ be a holomorphic self map of $\mathbb{D}$. Then C_ϕ is a bounded operator on $\mathrm{D}(\delta_1)$ exactly when*

$$M_\phi(w) = O\left(\frac{1-|w|^2}{|1-w|^2}\right) \qquad (|w| \to 1). \tag{3.3}$$

Proof: Suppose (3.3) holds. Using Theorem 1 we get

$$\|f \circ \phi\|_\bullet^2 = |f(\phi(0))|^2 + \int_{\phi(\mathbb{D})} |f'(w)|^2 M_\phi(w) dA(w) =$$

$$O\left(|f(\phi(0))|^2 + \int_\mathbb{D} |f'(w)|^2 \frac{1-|w|^2}{|1-w|^2} dA(w)\right) = O\left(\|f\|_\bullet^2\right) .$$

Assume now that C_ϕ is bounded. We have, using Lemma 2,

$$\|C_\phi f_p\|_\bullet^2 \geq \int_\mathbb{D} |f_p'(w)|^2 M_\phi(w) dA(w) =$$

$$\frac{|1-p|^2}{1-|p|^2} \int_\mathbb{D} |\alpha_p'(w)|^2 M_\phi(w) dA(w).$$

The change of variable $y = \alpha_p(w)$ together with the identity

$$M_{\alpha_p \circ \phi}(w) = M_\phi(\alpha_p(w))$$

give

$$\|C_\phi f_p\|_\bullet^2 \geq \frac{|1-p|^2}{1-|p|^2} \int_\mathbb{D} M_{\alpha_p \circ \phi}(y) dA(y) \geq$$

$$\frac{|1-p|^2}{1-|p|^2} \int_{r\mathbb{D}} M_{\alpha_p \circ \phi}(y) dA(y) \qquad (0 < r < 1).$$

As the integrand is harmonic we get

$$\|C_\phi f_p\|_\bullet^2 \geq 4 \frac{|1-p|^2}{1-|p|^2} M_\phi(p),$$

and (3.3) follows from Lemma 2 a). □

Recall that a bounded operator on a Hilbert space is compact when it maps weakly convergent sequences onto strongly convergent ones. It is easy to see that a sequence in $D(\delta_1)$ converges weakly to 0 if, and only if, it is bounded and covergent to 0 uniformly on compacta of $\mathbb{D}$. So we have the following criterion.

Lemma 3 *The following are equivalent:*

a) C_ϕ is compact on $D(\delta_1)$.

b) If (g_n) is a bounded sequence of elements of $D(\delta_1)$ with $g_n \to 0$ on compacta on $\mathbb{D}$, then $C_\phi g_n$ converge to 0 in $D(\delta_1)$.

Theorem 3 *Let ϕ be a holomorphic self map of $\mathbb{D}$. Then C_ϕ is compact on $D(\delta_1)$ if, and only if,*

$$M_\phi(w) = o\left(\frac{1-|w|^2}{|1-w|^2}\right) \qquad (|w| \to 1). \tag{3.4}$$

Proof:

Let (g_n) be a sequence in $D(\delta_1)$ such that $g_n \to 0$ on compacta on $\mathbb{D}$. Assume also

$$\|g_n\|_\bullet \leq 1 \quad \forall n \in \mathbb{N}.$$

Let $\varepsilon > 0$.

We can take $0 < r < 1$ such that

$$M_\phi(w) < \frac{\varepsilon}{3} \cdot \frac{1-|w|^2}{|1-w|^2} \quad \text{for } r \leq |w| < 1 \tag{3.5}$$

and $n_\varepsilon \in \mathbb{N}$ verifying

$$|g_n(z)|^2 < \frac{\varepsilon}{3}, |g'_n(z)|^2 < \frac{\varepsilon}{3} \text{ on } r\mathbb{D} \cup \{\phi(0)\} \text{ for } n > n_\varepsilon. \tag{3.6}$$

We have then, for $n > n_\varepsilon$,

$$\|C_\phi g_n\|_\bullet^2 = |g_n(\phi(0))|^2 + \int_{r\mathbb{D}} |f'_n(w)|^2 M_\phi(w) dA(w) \leq$$

$$\frac{\varepsilon}{3} + \frac{\varepsilon}{3} \int_{r\mathbb{D}} M_\phi(w) dA(w) + \frac{\varepsilon}{3} \int_{\mathbb{D}\setminus r\mathbb{D}} |g'_n(w)|^2 \frac{1-|w|^2}{|1-w|^2} dA(w) .$$

So, applying Theorem 1 to the identity function we get,

$$\|C_\phi g_n\|_\bullet^2 \leq \frac{\varepsilon}{3} + \frac{\varepsilon}{3} + \frac{\varepsilon}{3} \left(\|g_n\|_\bullet^2 - |g_n(\phi(0))|^2\right) \leq \varepsilon .$$

We conclude from Lemma 3 that C_ϕ is compact.

Conversely, assume that C_ϕ is a compact operator. We'll use the functions f_p and their properties listed in Lemma 2.

For any $p \in \mathbb{D}$, by Theorem 1, we have

$$\|C_\phi f_p\|_\bullet^2 \geq \int_\mathbb{D} |f'_p(w)|^2 M_\phi(w) dA(w) =$$

$$\frac{1-|p|^2}{|1-p|^2} \int_\mathbb{D} |\alpha'_p(w)|^2 M_\phi(w) dA(w) .$$

As $M_\phi(\alpha_p(w)) = M_{\alpha_p \circ \phi}(w)$ for $w \in \mathbb{D}$, we have

$$\|C_\phi f_p\|_\bullet^2 \geq \frac{1-|p|^2}{|1-p|^2} \int_\mathbb{D} M_{\alpha_p \circ \phi}(y) dA(y) \geq$$

$$\frac{1-|p|^2}{|1-p|^2}\int_{r\mathbb{D}} M_{\alpha_p\circ\phi}(y)dA(y) = 4\frac{1-|p|^2}{|1-p|^2}M_\phi(p) ;$$

where $0 < r < 1$.

By Lemma 2 we must have $\|C_\phi f_p\|_\bullet^2 \to 0$ as $|p| \to 1$, so

$$M_\phi(p) = o\left(\frac{1-|p|^2}{|1-p|^2}\right) \quad (|p| \to 1)$$

and the proof is complete. □

Bibliography

[1] A. Aleman. Hilbert spaces of analytic functions between the Hardy and the Dirichlet space. *Proceedings of the American Mathematical Society*, 115(1), may 1992.

[2] C. Carathéodory. *Theory of Functions*, volume 2. Chelsea, New York, 1954.

[3] L. de Branges and J. Rovnyak. *Square summable power series*. Holt, Rinehart and Winston, New York, 1966.

[4] J. Douglas. Solution of the problem of Plateau. *Trans. Amer. Math. Soc.*, 33:169–182, 1931.

[5] P. L. Duren. *Theory of* H^p *spaces*. Ac. Press, New York, 1970.

[6] John B. Garnett. *Bounded Analytic Functions*. Ac. Press, New York, 1981.

[7] J. E. Littlewood. On inequalities in the theory of functions. *Proc. London Math. Soc.*, 23:481–519, 1925.

[8] S. Richter. A formula for the local Dirichlet integral. *Michigan Math. J.*, pages 355–379, 1991.

[9] S. Richter. A representation theorem for cyclic analytic two-isometries. *Transactions of the American Mathematical Society*, 328(1), November 1991.

[10] D. Sarason. Doubly shift-invariant subspaces in H^2. *J. Operator Theory*, (16):75–97, 1986.

[11] Joel H. Shapiro. The essential norm of a composition operator. *Annals of Math.*, 125:375–404, 1987.

Bibliography

[12] Joel H. Shapiro. *Composition Operators and Classical Function Theory*. Springer-Verlag, 1993.

Notes

www.ingramcontent.com/pod-product-compliance
Lightning Source LLC
Chambersburg PA
CBHW030856180526
45163CB00004B/1593